Plants

Seeds

Patricia Whitehouse

Heinemann Library
Chicago, Illinois

Customer Service 888-454-2279
Visit our website at www.heinemannlibrary.com

Designed by Sue Emerson/Heinemann Library, Page layout by Carolee A. Biddle
Printed and bound in the U.S.A. by Lake Book

06 05 04 03 02
10 9 8 7 6 5 4 3 2 1

Library of Congress Cataloging-in-Publication Data
Whitehouse, Patricia, 1958-
    Seeds / Patricia Whitehouse.
        p. cm. — (Plants)
Includes index.
Summary: Introduces the physical traits, function, and uses of seeds.
    ISBN 1-58810-525-3 (HC), 1-58810-732-9 (Pbk.)
    1. Seeds--Juvenile literature. [1. Seeds.] I. Title. II. Plants
(Des Plaines, Ill.)
    QK661 .W45 2002
    5581.4'67—dc21

                            2001003651

Acknowledgments
The author and publishers are grateful to the following for permission to reproduce copyright material:
Title page, pp. 14, 16, 22T.L., 24T.L. E.R. Degginger Color Pic, Inc.; p. 4 Images International/Visuals Unlimited; pp. 5L, 23c Michael Gadomski/Bruce Coleman, Inc.; p. 5R Amor Montes de Oca; p. 6 Jerome Wexler/Visuals Unlimited; pp. 7, 15L, 21, 22B.L., 23d, 23e, 24B.L., Dwight Kuhn; p. 8, Mary Cummings/Visuals Unlimited; p. 9, 23a Tom Edwards/Visuals Unlimited; p. 10 Wally Eberhart/Visuals Unlimited; p. 11 David June; p. 12 Danny Camilli/Bruce Coleman Inc.; p. 13 Craig Mitchelldyer; p. 15R, 22R, 23b, 24R Walt Anderson/Visuals Unlimited; p. 17 Frank Lane Picture Agency/Corbis; p. 18 Rick Wetherbee; p. 19 Lynda Richardson/Corbis; p. 20 Rob and Ann Simpson

Cover photograph courtesy of Frank Lane Picture Agency/Corbis

Every effort has been made to contact copyright holders of any material reproduced in this book.
Any omissions will be rectified in subsequent printings if notice is given to the publisher.

Special thanks to our advisory panel for their help in the preparation of this book:

Eileen Day, Preschool Teacher
Chicago, IL

Paula Fischer, K–1 Teacher
Indianapolis, IN

Sandra Gilbert,
Library Media Specialist
Houston, TX

Angela Leeper,
Educational Consultant
North Carolina Department
of Public Instruction
Raleigh, NC

Pam McDonald, Reading Teacher
Winter Springs, FL

Melinda Murphy,
Library Media Specialist
Houston, TX

Helen Rosenberg, MLS
Chicago, IL

Anna Marie Varakin,
Reading Instructor
Western Maryland College

The publishers would also like to thank Anita Portugal, a master gardener at the Chicago Botanic Garden, for her help in reviewing the contents of this book for accuracy.

Some words are shown in bold, **like this.**
You can find them in the picture glossary on page 23.

Contents

What Are Seeds?

Seeds are part of a plant.

Some seeds are inside **fruits** and vegetables.

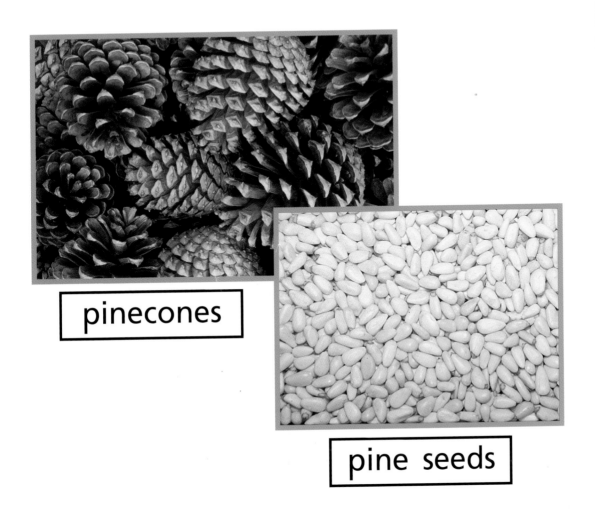

pinecones

pine seeds

Some seeds are inside **pinecones**.

The seeds come out when the pinecone opens up.

Why Do Plants Have Seeds?

Seeds make new plants.

The new plants look just like the plant they came from.

Where Do Seeds Come From?

Seeds come from plants.

The flower part of a plant makes seeds.

The flower part also makes a seed holder.

This is called the **fruit**.

How Big Are Seeds?

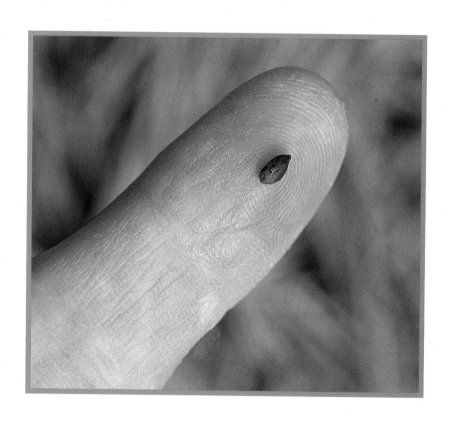

Seeds come in many sizes.

Lots of these flower seeds can fit on your fingertip.

Some seeds are very big.

This coconut is bigger than the girl's hands.

How Many Seeds Can Plants Have?

Some plants have just one seed.

This avocado has one seed.

Some plants have hundreds of seeds, like this dandelion.

Why Do Seeds Have Different Shapes?

seed points

Seeds have many different shapes.

Some have **points** that dig into dirt.

seed wings

seed hooks

Wings help some seeds blow in the wind.

Hooks help other seeds hold on to things.

What Colors Are Seeds?

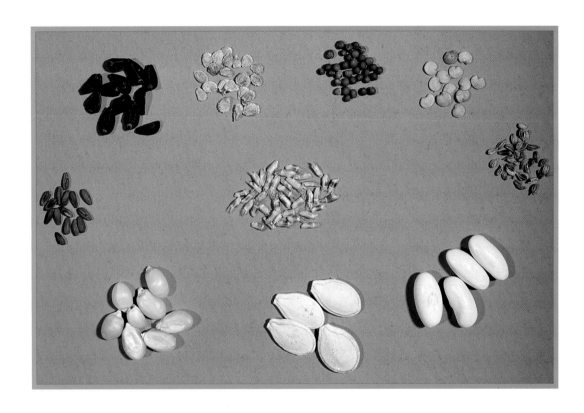

Most seeds are black, brown, or tan.

Some seeds have patterns on them.

These sunflower seeds have stripes.

How Do People Use Seeds?

People use seeds for food.

People eat some seeds just the way they are.

Some seeds are crushed, squeezed, or popped to make food.

People use seeds to grow new plants, too.

How Do Animals Use Seeds?

Animals use seeds for food, too.

Birds, squirrels, elephants, and monkeys eat seeds.

Some animals eat the seeds
right away.

Others save their seeds to
eat later.

Quiz

Can you remember what these seed parts do?

Look for the answers on page 24.

Picture Glossary

fruit
pages 4, 9

point
page 14

hook
page 15

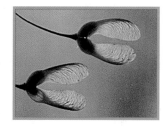
wing
page 15

pinecone
page 5

Note to Parents and Teachers

Reading for information is an important part of a child's literacy development. Learning begins with a question about something. Help children think of themselves as investigators and researchers by encouraging their questions about the world around them. Each chapter in this book begins with a question. Read the question together. Look at the pictures. Talk about what you think the answer might be. Then read the text to find out if your predictions were correct. Think of other questions you could ask about the topic, and discuss where you might find the answers. Assist children in using the picture glossary and the index to practice new vocabulary and research skills.

Index

Answers to quiz on page 22

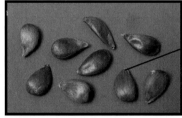

dig into dirt

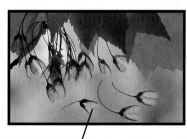

blow in the wind

hold on to things

24

PRIMARY SOURCES OF
FAMOUS PEOPLE IN AMERICAN HISTORY™

ANNIE OAKLEY

WILD WEST SHARPSHOOTER

JASON PORTERFIELD

rosen central
Primary Source™
The Rosen Publishing Group, Inc., New York

Published in 2004 by The Rosen Publishing Group, Inc.
29 East 21st Street, New York, NY 10010

Library of Congress Cataloging-in-Publication Data

Porterfield, Jason.
Annie Oakley: Wild West sharpshooter/ Jason Porterfield.— 1st ed.
   v. cm. — (Primary sources of famous people in American history)
Includes bibliographical references (p.   ) and index.
Contents: Annie's early life—A contest with Frank Butler—Buffalo Bill's Wild West
show—A world star—The legend of Annie Oakley.
ISBN 0-8239-4102-7 (lib. bdg.)
ISBN 0-8239-4174-4 (pbk.)
6-pack ISBN 0-8239-4301-1
1. Oakley, Annie, 1860-1926—Juvenile literature. 2. Shooters of firearms—United
States—Biography—Juvenile literature. 3. Women entertainers—United States—
Biography—Juvenile literature. 4. Frontier and pioneer life—West (U.S.)—Juvenile
literature. [1. Oakley, Annie, 1860-1926. 2. Sharpshooters. 3. Entertainers. 4. Women—
Biography.] I. Title. II. Series.
GV1157.O3L56 2003
799.3'092—dc21

2003003809

Manufactured in the United States of America

Photo credits: cover © Hulton/Archive/Getty Images; pp. 4, 5, 9, 11, 13, 16, 29 courtesy of Garst Museum;
pp. 7, 28 Ohio Historical Society; p. 8 David Rumsey Historical Map Collection, www.davidrumsey.com; pp. 10
(Otto Westerman, X-6484), 15 (NS-664), 18 (X-31721), 19 (NS-456), 25 (NS-150) Denver Public Library, Western
History Collection; p. 12 © Culver Pictures; p. 14 Rare Book, Manuscript, and Special Collections Library, Duke
University; p. 17 Western History Collections, University of Oklahoma Library; p. 20 Buffalo Bill Historical
Center, Cody, Wyoming; 1.69.1070; pp. 21, 24 © Bettmann/Corbis; p. 23 Circus World Museum, Baraboo,
Wisconsin; p. 27 Library of Congress Prints and Photographs Division

Designer: Thomas Forget; Editor: Jill Jarnow; Photo Researcher: Rebecca Anguin-Cohen

L-3.4/P0.5

CONTENTS

Annie Oakley was born on August 13, 1860. Her parents were Jacob and Susan Moses. They named their new little girl Phoebe Ann. Everyone called her Annie.

The Moses family lived in Ohio. They owned a farm in Darke County. The whole family helped on the farm. Annie's father hunted and trapped animals for food.

Phoebe Ann Moses was born in this house in 1860 in Greenville, Ohio. Phoebe Ann later changed her name to Annie Oakley.

Annie was the fifth daughter of Jacob and Susan Moses. Jacob Moses died in 1866.

Jacob Moses died when Annie was only six years old. Susan Moses sold the farm to pay off debts. She moved her children onto a rented farm. Life on the new farm was hard. Annie and her brother John helped the family survive by hunting.

ANNIE'S BIG FAMILY

Annie had eight brothers and sisters! Her mother was married several times.

Here is an Ohio farm photographed between 1886 and 1888. The Moses family farm may have looked like this.

Annie left home to work when she was ten. She lived at a nearby infirmary. Sick people and orphans lived there. Annie was paid to sew and to take care of children.

Annie returned to the farm and began to hunt again. She became the best shot in Darke County. The Moseses ate many of the animals she shot. She sold the rest to stores.

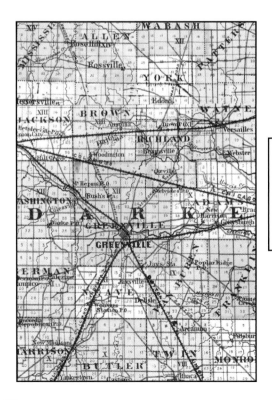

Drawn in 1875, this map of Ohio shows Greenville in Darke County where Annie Oakley was born.

Annie Oakley sold some of her game to the G. A. Katzenberger & Brothers Family Grocery store in Greenville, Ohio. The store is shown here in the early 1800s.

2 A CONTEST WITH FRANK BUTLER

Annie's hunting helped get her family out of debt. She never missed a shot. Annie's friends talked her into entering shooting contests.

People at the time used guns to hunt for food and to protect themselves. Many felt that women should not hunt or shoot guns. Using a gun was considered to be unladylike.

Annie helped her family by shooting wild animals. They ate what she killed. She sold the rest to storekeepers.

In 1892, Annie went to London. She performed for Queen Victoria. This photo of Annie was taken during that trip.

Annie often amazed men with her sharpshooting. She won many matches.

In the spring of 1875, Annie entered a shooting match in Ohio. She shot against a man named Frank Butler. Frank was one of the best shots in the country.

Annie beat Frank and they fell in love. Frank and Annie married in June 1882.

Annie was one of the most admired women of her day. Smart, pretty, and gentle, she became famous in a sport that was reserved for men.

Frank Butler was amazed when Annie beat him in a shooting match in 1875. This photo of Frank is from 1880.

3 BUFFALO BILL'S WILD WEST SHOW

After they got married, Annie and Frank worked as a team. Annie changed her last name to Oakley. She thought her new name had a stronger sound.

The team of Oakley and Butler toured the country. They shot in matches and gave demonstrations. Annie soon took over the act. Frank became her manager and stage helper.

A collector's card from a cigarette pack shows Miss Annie Oakley. It pictures other famous sharpshooters. They are Captain A. H. Bogardus, Hon. W. F. Cody (Buffalo Bill), and Dr. W. F. "Doc" Carver.

This picture of Annie dates from about 1901. She was practicing her jumps. During Buffalo Bill's Wild West show, she jumped over a rifle.

Oakley and Butler performed throughout America. They made many friends as Annie became famous.

The Sioux Indian chief Sitting Bull saw Annie's act. Impressed by her sharpshooting, he adopted Annie as his daughter. He named her Watanya Cecilla. The name means "Little Sure Shot" in Sioux.

Annie and Frank's dog, Dave, even had a part in the act. Annie shot apples off of his head. Dave would catch the pieces in his mouth!

Annie and the Indians pose in London on a set of fake mountains. Buffalo Bill's Wild West performed for Queen Victoria at her 1887 Golden Jubilee.

Buffalo Bill Cody asked Annie and Frank to join his show. They became part of Buffalo Bill's Wild West in 1885.

They toured all over. Cowboys, sharp-shooters, and Native Americans showed people what life was like in the West. Annie performed in front of huge crowds. Sitting Bull was in the show, too.

Chief Sitting Bull, shown here, called Annie "my daughter, Little Sure Shot." Annie taught Sitting Bull to read and write.

Annie takes shooting practice in 1892. She is in Earl's Court, London.

4 A WORLD STAR

The Wild West show made Annie Oakley even more famous. She practiced a lot.

People came just to watch Annie shoot. She shot cigarettes from Frank's mouth. Frank threw glass balls up in the air for Annie to shoot. Her most famous trick was to shoot behind her back. She aimed through a mirror and hit the target every time!

Annie learned to sew when she was young. Later she sewed her own costumes. She packed them neatly for traveling in this special trunk.

One of Annie's most famous tricks was to shoot backward by looking into a mirror. Here she poses for a picture.

The Wild West show went to Europe in 1887. Annie won many gold medals. Fans sent her gifts. Famous people went to the shows to see Annie shoot. She even won a match against Grand Duke Michael of Russia!

SHOOTING FOR THE QUEEN

Great Britain's Queen Victoria wanted to see Annie shoot. She watched the show from a box seat covered in velvet drapes.

This poster highlights scenes from Annie's life as a performer. In addition to being a sharpshooter, she was a fine horsewoman.

Annie showed the world that women could shoot as well as men. Annie began teaching women how to shoot. She felt that they should be able to defend themselves.

Her work was not all about shooting. She also worked for other causes. Annie gave much of her money to charity. She also spoke out for Native American rights.

Annie believed women should know how to shoot guns. In this 1918 photo, she is teaching a group of women in North Carolina.

Buffalo Bill stands with his troupe in Rome, Italy, in 1890. Annie is in the second row on the right. Frank Butler is at the right end of the same row.

5 THE LEGEND OF ANNIE OAKLEY

Annie was badly hurt in a train wreck in 1901. She finally had to leave Buffalo Bill's Wild West. She and Frank thought about retiring.

But Annie became restless. She joined the Young Buffalo's Wild West in 1911. More people than ever came to see Annie perform. She and Frank left the show after two years.

ANNIE'S WAR EFFORT

In 1917, Annie offered to train a regiment of women to fight in World War I. President Woodrow Wilson never responded to her offer.

Annie poses with a gun given to her by Buffalo Bill. The year is 1922.

Annie and Frank kept busy until 1926. They both fell ill that year. Annie died on November 3, 1926. Frank died 18 days later. They were buried together in Greenville, Ohio.

Annie Oakley's memory lives on. We see her in books, movies, and plays. We will remember Annie for her shooting skill and her thoughtful, exciting life.

ANNIE OAKLEY
1926
AT REST

Annie died in 1926. Frank died 18 days later. They are buried side by side in Brock Cemetery, Greenville, Darke County, Ohio.

Annie Oakley, Frank Butler, and their dog, Dave, worked at the Carolina Hotel in North Carolina. They performed and taught people how to shoot. Dave the dog was also in the act.

TIMELINE

1860—Annie Oakley is born on August 13.

1882—Annie marries Frank Butler.

1885—Frank and Annie join Buffalo Bill's Wild West show.

1887—Annie performs for Queen Victoria while touring Europe.

1901—Annie is injured in a train wreck.

1926—Annie dies on November 3 and is buried in Greenville, Ohio.

GLOSSARY

adopt (uh-DOPT) To raise a child of other parents.

cause (CAWZ) An idea that a person supports.

debt (DET) Something owed.

demonstration (DEH-mun-STRAY-shun) Showing people how to do something by acting it out.

game (GAYM) The meat of wild animals that are hunted for food; wild animals that are hunted for food.

infirmary (in-FUR-muh-ree) A place for the caring of the sick and needy.

manager (MAN-uh-jur) Someone in charge of a store, business, etc., or in charge of a group of people at work.

match (MACH) A game or contest in which two or more people compete for a prize.

orphan (OR-fuhn) A child or animal who no longer has parents.

skill (SKIHL) The level at which a person performs.

tour (TOOR) To travel for the purpose of giving shows.

unladylike (un-LAY-dee-LIKE) Poor behavior for a lady.

WEB SITES

Due to the changing nature of Internet links, the Rosen Publishing Group, Inc., has developed an online list of Web sites related to the subject of this book. This site is updated regularly. Please use this link to access the list:

http:// www.rosenlinks.com/fpah/aoak

PRIMARY SOURCE IMAGE LIST

Page 5: Annie Moses Butler, about 1880. Photograph by Martin, Chicago, housed in the Darke County Historical Society's Garst Museum.

Page 8: *Cram's Rail Road and Township Map of Ohio*, 1875, by George F. Cram.

Page 9: G.A. Katzenberger & Brothers Family Grocery store, photograph, circa 1880, Greenville, Ohio, courtesy of Garst Museum, Darke County Historical Society, Greenville, Ohio.

Page 10: Annie as purveyor of game, photograph, Ohio Historical Society, Columbus, Ohio.

Page 11: Annie Oakley in London, photograph, 1892, Darke County Historical Society, Garst Museum, Greenville, Ohio.

Page 12: Formal portrait of Annie Oakley, photograph, Annie Oakley Foundation.

Page 13: Frank Butler, photograph, about 1880 by Martin, Chicago, Darke County Historical Society, the Garst Museum, Greenville, Ohio.

INDEX

ABOUT THE AUTHOR

Jason Porterfield is a writer living in Chicago, Illinois.